ISBN: 978-1-7382643-1-5

c·o·u·l·d

City · Of · Unintentional · Littering · Debris

verb

- *We could see something like this coming.*
- *A lot of urban runoff could be prevented.*

Litter and debris in our environment remind us of the unintentional nature of certain events in a place known as THE COULD.

Hello Visitors,
Welcome to the COULD!

We, the people of this town, also known as The City of Unintentional Littering and Debris (The COULD), want to address the issue of littering and debris. We are committed to keeping our town clean and taking responsibility for our actions. We regret the negative impact of our unintentional actions and are dedicated to creating a cleaner environment. We can all do better and be more mindful of practicing waste disposal etiquette.

Thank you for your contribution,

The Mayor's Office

100 C.O.U.L.D, Drive
City Hall Place

Get ready for lots of rain tonight and tomorrow morning.
[CC]: **STORM ALERT**: Expect a 24-hour storm with heavy Rain and Wind. Once the Storm passes, however, we are headed into Sunny Weather.

Okay, it's bedtime. Your science project looks fantastic and your teacher will be amazed that it's an A+
SCIENCE PROJECT PAPER

I am thankful that I'll have the chance to showcase my science project tomorrow!

SCHOOL BUS

The rain destroyed my Science Project. I didn't mean for it to happen.
SCIENCE PROJECT PAPER

Rain + Leaching = What Are Its Effects?

Leaching is the process of contaminants, such as water-soluble paint, pesticides or fertilizers, being carried by water downward through permeable soils and passing through street drains. This can lead to the pollution of streams, lakes, and drinking water reservoirs. It can also affect the marine ecosystem where the fish we eat live. Studies have shown the presence of pesticide compounds in 89% of sampled streams (Romanok, et al 2021), (PES)

It's important to recognize that some events and circumstances can unintentionally lead to waste generation. For instance, when someone is rushing to catch a bus, their critical personal belongings might accidentally fly out of their bag. What was once important to them is now considered waste, which may not be found and could end up as urban runoff on the next windy, rainy day. This highlights the need for preventive measures to avoid such unintentional waste generation.

Oh No!!!

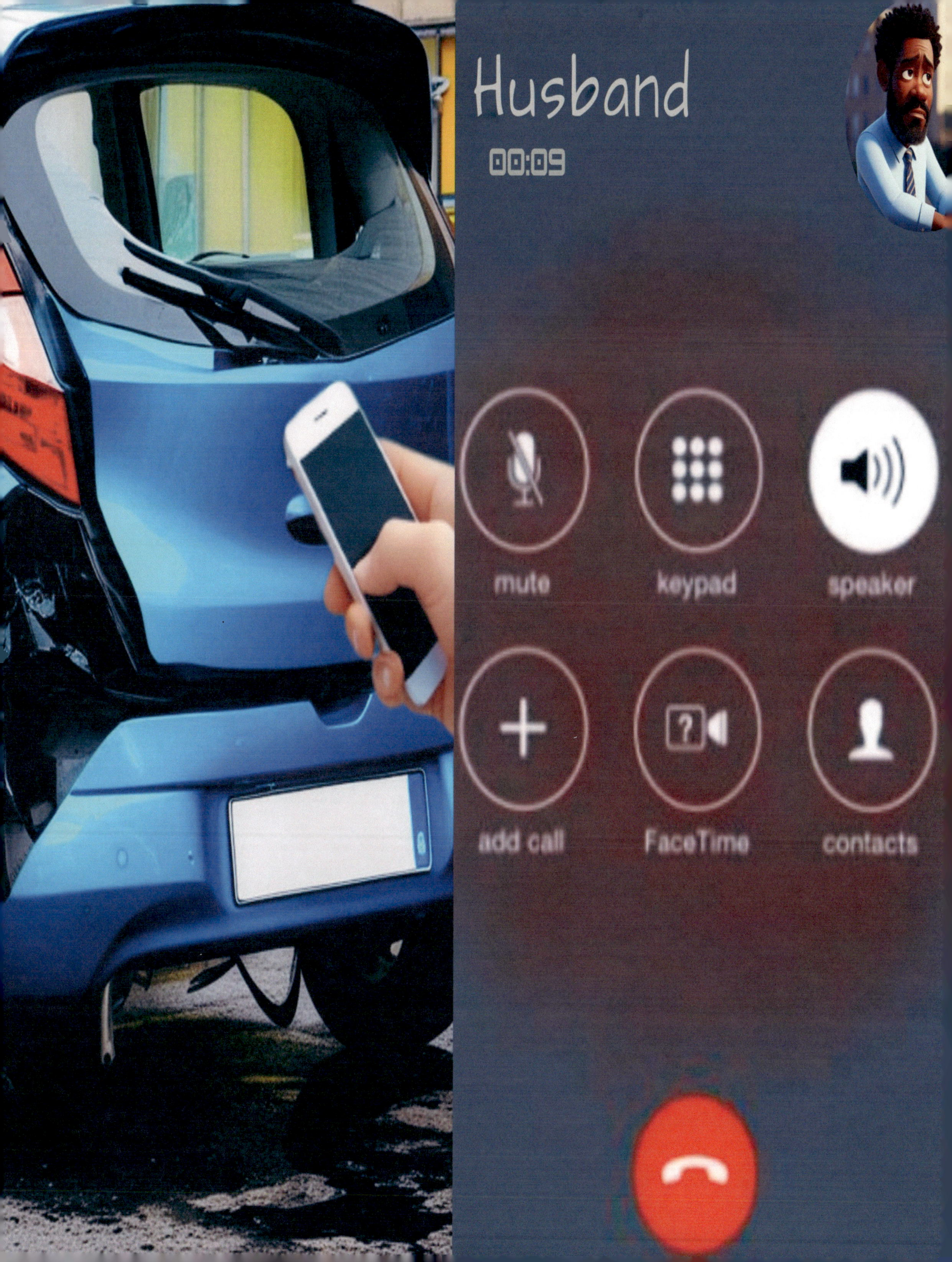
Husband
00:09
mute
keypad
speaker
add call
FaceTime
contacts

Please Please
PICK UP!

This Is Draining $!$!$!

[CC]: It's going to be a hot day. The weather will be HOT, with a high of 33°C / 92°F and clear skies all day and night.

WASTE
BIN

On a sunny day, Mom takes her children for a walk. The younger child accidentally dropped her hand-held clear glitter toy while the mom was focused on her son's rollerblading. The twins were unable to communicate, and the mom, focusing on her son, who is rollerblading, was unable to recognize the unintentional dropping of the sparkle glitter toys (aka microplastics).

With their increasing energy use and pollution throughout their life cycle, single-use face masks have become a pressing environmental issue. Accidentally discarded masks contribute to plastic pollution and pose a threat to wildlife. This urgent problem requires our immediate attention and action to prevent further harm (Wang, 2023).

- What are the most effective waste disposal habits that we can promote to curb the unintentional littering of single-use face masks in our cities, towns, and villages?

- What strategies can be implemented to encourage responsible debris disposal and reduce littering?

Each of us has a role to play in addressing this issue, and together, we can reduce the environmental impact of single-use masks and other littered debris.

Tire Rubber: Cars, Bicycles, Scooters, Carts, and e-bikes

Approximately 3 Billion new tires are manufactured annually, but a distressing 800 million are discarded annually. The pollutants released from tire wear significantly threaten the environment and human health. Cars, Bicycles, Scooters, Carts on wheels, and e-bikes, to name a few: As tires wear down, the worn-off tire pieces release pollutants into the air, water, and soil, with far-reaching effects on various ecosystems (Mayer, 2024).

THE COULD
Car Free Day

Car-free day provides a convenient, enjoyable, and affordable way to explore the city, support local businesses, and travel sustainably - Mayor's Office.

STREET SWEEPING

STREET SWEEPING

What The Street Sweepers Miss, The Magnify Glass Exposes.
Micro - Nano Plastic, Sediment On Roads

Seabirds With Their Innocent Curiosity

Curious seabirds mistook a brown paper bag for food and scattered its contents—microbeads intended for jewelry making—into the ocean, unintentionally polluting the marine ecosystem. They successfully stole food from humans 50% of the time, which is higher than their success with other species. Growing interactions between humans and wildlife have worsened litter pollution (Raghav & Boogert, 2022).

Unintentionally Scattering Urban Litter Across The Water's Surface

The birds eagerly scatter urban litter across the water's surface, unaware of the consequences. While flying, they believe they are bringing food back to the nest, only to discover that they have been collecting plastic pieces instead of actual food.

WHAT COULD BE?

We must reduce unintentional litter and urban runoff to foster a sustainable future for The [city] COULD. Recent data shows that much waste is likely unintentional; for example, the frequent presence of public transit tickets in runoff suggests that many individuals are using public transport, a positive step in reducing our carbon footprint. THE COULD relies on its citizens to promote proper waste disposal etiquette. A simple reminder can encourage mindfulness, as many littering instances are accidental rather than deliberate.

A+
COULD BE ACCOMPLISHED

It will rain, and accidents will happen. We might not be able to control the unintentional, but as individuals, we can capture and reduce the potential impacts of our actions. We can take personal responsibility to reduce the likelihood of accidental littering and its environmental consequences. Watching the rain and wind strip the paint and objects from the student's science project, we can not help but see the avoidable negative ecological impact. The student's unintentional letter failing grade of F on the science project is a lesson learned. But it is not the end. We can use these experiences to highlight the potential for improvement to prevent similar incidents and, more importantly, to secure an A+ on the next science fair project for hard-working students, which could benefit the classroom and, ultimately, the marine ecosystem.

THE C.O.U.L.D REFERENCES

Kamand, Jean-Paul, Leaf Litter, Its Adhesive and Camouflaging Characteristics Have a Striking Effect on the Eco-System (June 1, 2023). Available at SSRN: https://ssrn.com/abstract=4461616 or http://dx.doi.org/10.2139/ssrn.4461616

Paul M. Mayer,Kelly D. Moran,Ezra L. Miller,Susanne M. Brander,Stacey Harper,Manuel Garcia-Jaramillo,Victor Carrasco-Navarro,Kay T. Ho,Robert M. Burgess,Leah M. et al.

Where the rubber meets the road: Emerging environmental impacts of tire wear particles and their chemical cocktails. Publication: Science of The Total Environment, Publisher: Elsevier. 1 June 2024

Pesticide Environmental Stewardship (PES) The Problem of Leaching Retrieved from:https://pesticidestewardship.org/water/leaching/

Raghav, S., & Boogert, N. J. (2022). Factors associated with Herring Gulls Larus argentatus stealing food from humans in coastal towns. Bird Study, 69(3–4), 103–108. https://doi.org/10.1080/00063657.2022.2162846

Romanok, K.M., Bradley, P.M., and Mahler, B.J., 2021, Pesticide and transformation product concentrations and risk quotients in U.S. headwater streams: U.S. Geological Survey data release, https://doi.org/10.5066/P9PSU28Y.

Wang, Limin et al. "Global face mask pollution: threats to the environment and wildlife, and potential solutions." The Science of the total environment vol. 887 (2023): 164055. doi:10.1016/j.scitotenv.2023.164055

Made in the USA
Las Vegas, NV
07 November 2026

57672511R00024